Railways Around Leicester

Scenes of Times Past

Above: A local train to Peterborough, here in 1962 hauled by Britannia, *leaves from platform 1 the site of the former Midland Counties Railway Campbell Street station building.*

Previous page: Having just left Regent Road bridge in its wake in August 1961 a Leicester summer Sunday excursion train heads for Southend-on-Sea.

Railways Around Leicester

Scenes of Times Past

by
HORACE A GAMBLE

Anderson Publications

To all those lads
who, like me, were educated
on the platforms of London Road Station
in the 1930s

First published in 1989

BRITISH LIBRARY CATALOGUING-IN-PUBLICATION DATA

Gamble, Horace A. (Horace Alfred
Railways around Leicester: scenes of times past.
1. Leicestershire. Leicester. Railway services, history
I. Title
385'.09425' 42

ISBN 0 907917 11 9

Printed in Great Britain by
AB Printers Limited for
Anderson Publications, 29 The Fairway,
Blaby, Leicester LE8 3EN

Author's Acknowledgements
I wish to thank most sincerely Mr Colin Garratt
for writing the Foreword, and Messrs P.H. Anderson
and S.J. Dixey for their assistance
most willingly given.

Contents

Seventy years after Knighton new tunnel was built in 1893 one of the standard LMS heavy freight locomotives, 48170, threads the bore. Many of these saw War Department service in the Middle East.

Foreword

Horace Gamble — the archetype enthusiast — was the founder of the Leicester Railways Society which for many years provided a focal point for our enthusiasms. Railways were a passionate and serious business to those who thronged the Society's clubroom. The highlight of the Society year was W A Camwell's movies. These were hardcore films for the connoisseur; they dated back to the 1940s and the diversity and rariety of many of the scenes reduced us to a stupor as we sat in the clammy darkness.

For all its local interest, this book is but a microcosm of a once ubiquitous scene; virtually every part of the country had railways whose devotees expressed their joys much as Horace has done. Train spotting was the principal manifestation of railway interest and there was no finer activity for a boy to indulge in. The railway was Britain's premier industry, the pulse of the nation and even the broadest understanding of it brought myriad benefits some of which can only be obtained today from the very finest of private educations. If one understood the railway one understood the country; geographically, industrially, socially and commercially. Locomotives with their numerous designs sharpened the eye for detail and the mind for aesthetics. Numbers and dimensions opened the mind to statistics whilst travel, which is inherent to train spotting, broadened the mind and heightened perceptions. The vastness and absolutness of the railway commanded respect and engendered discipline; it was the perfect basis for moulding young minds.

Those of us privileged to have lived through the golden years embraced by this book crave for them still but the world moves on though not always benificially. Sadly railways are no exception and nowadays when I take my young son into the hallowed portals of Leicester London Road station I realise that there is no way of conveying to him the magic of the past. Yet the ghosts still haunt; the brickwork still echoes to excited cries of school friends and their faces emerge from sublimal memory as fresh and youthful as in their adolescent prime. Standing there in the modernity of today I look into James' eyes and see only the light of the years to come shining from them and realise that it is all a dream; a dream of the greatest legend of all time.

Colin Garratt

Newton Harcourt
1989

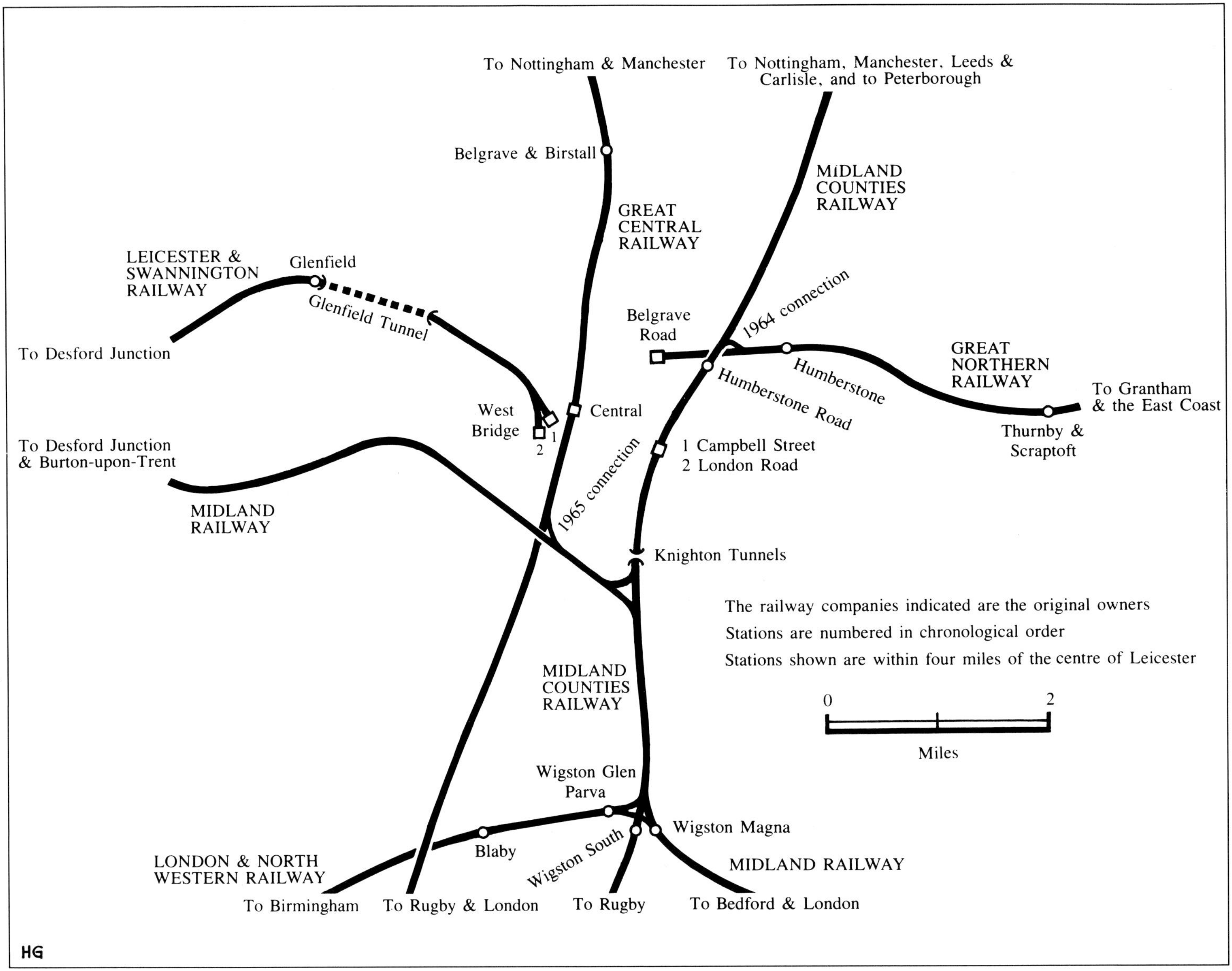

Railways in Leicester and its Environs

The Midland Line

Becoming a railway enthusiast, with its attendant pleasures and consuming passions, was in no small measure due to the indirect influence of my grandfather. For as an employee of the Midland Railway he lived in a company house in Slate Street overlooking London Road station at Leicester and from which I spent much of my formative years watching trains go by.

It was not that grandpa was an engine man. Far from it, he did not know one locomotive from the next. His speciality was cart horses not iron horses. Indeed if ever he had a dislike of locomotives it was well founded for he had a very narrow escape from death when almost run down in the dark by a goods train. Dropping flat onto the sleepers he lay there quaking as it clanged overhead. Next morning when he looked into his shaving mirror he saw a different man for overnight his hair had turned white.

His early years in railway service were spent as a shunter in Chaddesden sidings, at Derby, when horses were still used to move wagons. Then at about the turn of the century he came to Leicester, living in Knighton Fields, and working at Knighton sidings. After failing an eyesight test during an interview for the job of goods train guard he transferred to the cartage department at Queen Street goods depot and moved into the house in Slate Street.

There were the times when he went away to weekend conferences and then I would be invited to stay with my grandma; the greatest treat imaginable. In the evening after I had been put to bed and the sputtering gas lamp in the bedroom had been extinguished the ceiling would be illuminated by the lights of the station, occasionally obliterated or dappled by flitting shadows of scudding steam as a goods train laboured past. Far from disturbing my rest the sounds of trains in the night were music to the ear and stirring to the soul.

It was not until I was older that I learned of the reason for grandpa's trips away. Although he had little formal education he held several high positions in railway associated service. Among these were president of the National Foreign Free Pass Movement, three times chairman of the Leicester and District Committee of the Railway Employees Privilege Ticket Association, and finally chairman of the Association itself.

During the day my field of vision from a wooden crate in grandpa's back yard took in the up and down goods lines, much of platform 4 and the carriage sidings. Light engine movements between the locomotive sheds and the southern end of the station were made by traversing the goods roads. At that time a former Midland Railway passenger tender engine always presided over the carriage sidings and seemed to spend much of the day doing little

LEICESTER (CAMPBELL STREET) STATION 1840–92

Leicester's second station was designed by local architect William Parsons. Described as 'a fine large classical building eleven bays wide and with a giant portico of six Tuscan columns carrying a pediment', its clock was illuminated by gas. The first floor was devoted to the Board room and head offices of the Midland Counties Railway.

but simmer, while the crew sat reading newspapers.

Opposite the bay-windowed Leicester East signal box on the station there was a crossover from the siding nearest platform 4 by which milk tanks or horse boxes could be shunted onto the rear of waiting local trains. My favourite sighting however was the daily train from Yarmouth to Birmingham which arrived around one o'clock with a Midland & Great Northern Joint Railway locomotive in glorious yellow-ochre livery. After the engine was uncoupled it repaired to the sheds to turn, take coal and water before returning on the corresponding down train in mid afternoon.

The mysterious ritual of preparing a locomotive for the return journey in the shed yard was something with which I later became familiar from another seat of learning, that excellent alleyway raised on stilts above the railway, which runs from Swain Street to Hutchinson Street and commonly known as the Birdcage, now officially Hutchinson Walk. It was built in 1893 at a time when great improvements were taking place on Midland Railway installations at Leicester. Some years elapsed before I was allowed to venture there on my own. However an elderly neighbour would often take my brother and I on her weekly expedition to the local shops whereupon we would demand that the return walk home must be routed via the Birdcage.

My first acquaintance with railway literature was gained in that modest house in Slate Street for every month the appearance of grandpa's copy of the *LMS Magazine,* price one penny, would be eagerly awaited. Although it contained, among the commercial and

London Road station frontage in 1973. The Midland Railway's architect, Charles Trubshaw, has in this building reflected the Victorian self-assurance of a strong and confident company. Whilst undeniably striking, with a decorated orange-brown terra-cotta facing, it lacks the gracious elegance of Campbell Street station which it replaced in 1892.

technical items, very little locomotive topics it never failed to fascinate and to whet the appetite. Eventually each copy would be passed on to me and cut up to form engine picture scrapbooks.

It was always a novelty to be able to go on to the station officially without payment. This sometimes happened when grandpa wished to pass some official papers connected with one or other of his committees to a certain gentleman in the telegraph office. I would be asked to act as his courier; a request I never refused!

Concurrently with grandfather's retirement after forty years service, and his departure from Leicester, my parents made a move from our home, also in Slate Street, to an address off Evington Road. From here one could still hear the trains in the distance at night after the rest of the daily hubbub had subsided. New Year's Eve is especially remembered for the highly audible clamorous greeting sent out at midnight by church bells, factory sirens, and the shrill whistles and gruff hooters of steam locomotives in the shed yard.

Whilst a pupil of Alderman Newton's Boys School the Birdcage became my second home. Here especially on a Saturday, would congregate a motley group of schoolboy engine fans, writing down locomotive numbers in grubby Woolworth's notebooks. Some of the lads whose friendship was conjoined in those days were still active enthusiasts fifty years on.

Nowadays it is hard to visualise the exciting panorama of the engine shed complex one could see from this magnificent vantage point. Although the Midland Counties Railway had been built

The last two remaining houses in Slate Street numbers 16 and 18, left, thirty years after the latter had been vacated by my grandfather. Before then it had been my watchtower for all that went on in these sidings as well as on platform 4 and along the goods roads. When this photograph was taken in 1966 the Midland Railway signals were about to be replaced by upper quadrants after more than seventy years service.

The Midland Railway coat of arms on the parapet of the bridge carrying London Road over the tracks. The shield bears six heraldic devices denoting places embraced by the early Midland Railway and is surmounted by a Wyvern, the emblem of the Leicester & Swannington Railway. Birmingham, Leeds, Bristol, Leicester, Lincoln and Derby were represented.

through Leicester almost one hundred years previously the original locomotive round shed, topped by a weathervane bearing a two dimensional locomotive usually thought to be a facsimile Jenny Lind type, was still in use, to house and service small tank engines.

Next to it was a similar but larger roundhouse built a little later by the Midland Railway and inhabited by freight tender locomotives. The third, a straight shed, was erected in 1893 for housing passenger engines. In its heyday the depot had one hundred locomotives allocated to it and in the late 1930s still hosted over eighty. Rebuilding was long overdue. The coaling of locomotives was still a most primitive affair done manually by craning in square wooden tubs the coal shovelled from the forty-ton bogie wagons specifically built and lettered Loco Coal. At the rear of the coal stage was a corrugated iron hut which housed the wheel drop; two old sheerleg cranes were used to lift locomotives when bogies needed to be removed. Scattered around the yard were many water cranes, and gas lamp posts. A hand operated turntable completed the scene, whilst in the background, stood the handsome and extensive Queen Street goods warehouse.

Railways entered into my home life as a hobby with clockwork trains and over the years we built up quite a sizeable layout in the attic. Hand in hand with Hornby trains went the purchase monthly of the *Meccano Magazine* containing articles on model railway practice as well as news and features on main line railways. Later

The original Midland Counties Railway engine roundhouse contrasts piquantly, in this 1938 scene, with three newish locomotives. The two passenger tank engines 178 and 182 were only a few months old, and, the heavy freight locomotive 8030 sandwiched between them had been built in 1936. It survived hostilities in War Department service and was sold to the Iranian State Railways in 1948.

I felt really grown up when purchasing my first copy of the monthly *Railway Magazine* price one shilling, a periodical I bought continuously for almost fifty years.

About this time I first handled a camera, borrowed on occasion from dad. My parsimonious pocket money was measured in pence so that films and their subsequent processing were a rare luxury. My early attempts at train photography, some taken with the aid of a bicycle crossbar on which to stand to gain the requisite height, were in the main abysmal.

When my schooling was completed I became an apprentice draughtsman with a company in South Wigston which rewarded me with a weekly wage of fifteen shillings. This opened up a new world. Not only was it possible to save up the train fare for trips to places such as Nuneaton, Rugby and Birmingham but I could travel to work daily by train using any one of the three stations in Wigston. A weekly season ticket valid between Leicester and all of the Wigston stations was two shillings and ninepence, and monthly Saturday evening excursions from Leicester to Birmingham cost one shilling and seven pence return.

Permission could be obtained in advance from the railway company to visit nominated engine sheds provided that rail tickets were shown on arrival to prove that one had travelled there by

A familiar scene, especially during summer months, was the rostering of nominally goods locomotives for holiday traffic. Here, in July 1961, at London Road station 44465 moves the empty coaching stock of a train from Sheringham out of platform 3 en route for stabling at Wigston carriage sidings.

train. For such a purpose a group of like-minded friends often would take advantage of the Birmingham evening excursions. There we revelled in unfamiliar railway sights and sounds especially at the magnificent Snow Hill station of the Great Western Railway. Just before World War II commenced, and railway excursions were withdrawn, we made our last evening trip, this time to Manchester, Victoria, at a return fare of two shillings and three pence! The hardcore of Leicester enthusiasts who took part in these outings formed themselves into the Leicester Railway Society but it was not until the war had ended that it grew and flourished.

The largest volume of traffic which passed in and out of Leicester was on the London Midland and Scottish Railway route through London Road station so it was natural that the fledgling Society spent much time meeting on its platforms. On cold nights there was always the refreshment room to provide hot cups of tea and, usually, a coal fire roared in the waiting room to welcome us when lulls occurred in the timetable. The large mahogany table was often the pitch for exuberant games of shove ha'penny much to the disgust of bona-fide travellers.

Many a time we were allowed on to the hallowed environment of the footplate and shown the art of placing shovelsfull of coal in

Following nationalisation of the railway companies in 1948, and the subsequent locomotive exchanges, a Cleethorpes to Birmingham service, extended to Sidmouth in summer months, was instituted. This brought Eastern Region B1 class locomotives, so common at Leicester Central, into London Road for the first time. Here the return train about to depart in August 1962 is headed by B1 61143. Approaching the station is a Melton Mowbray to Leicester stopping train with 42334, a Leicester stalwart of over thirty years standing.

exactly the right spot in the firebox to produce a successful steam-raising fire. We got to know many of the drivers by name and remembered their duty rosters so that we could always be sure of hearing some interesting yarns when next they were working a particular train.

Prior to the erection of London Road station the Midland Counties Railway's Campbell Street station covered approximately the same site. However its main buildings were at rail level with the forecourt accessible from both Campbell Street and Station Street. Today two bulky stone pillars at the end of the latter are the sole remains.

The station was built in 1840 on the then main route from York to London, Euston, by way of Wigston, Rugby and the London & Birmingham Railway. Leicester was the headquarters of the Midland Counties Railway until 1844 when the Company amalgamated with the North Midland Railway and the Birmingham & Derby Junction Railway to form the Midland Railway, with head offices at Derby. Increasing traffic and subsequent delays at Rugby caused the Midland Railway to open a new line from Wigston North Junction to Hitchin in 1857 thus gaining access to London, Kings Cross, by way of Great Northern Railway. At that time an additional platform was built at Leicester and this in turn was converted into an island platform in 1869 after the Midland Railway had completed its own route into London, at St Pancras, the previous year.

London Road station was built 1892-3 following the demolition of Campbell Street station. Its two island platforms were longer than the original necessitating the removal of Upper Fox Street bridge and the building of Swain Street bridge. This was provided with a parallel footbridge and stairs giving direct access to the platforms. London Road was then an open station, as had been Campbell Street, and tickets were collected at platforms remote from the main station. From 1892 onwards the northbound ticket platform was situated between Knighton tunnel and Welford Road bridge, whilst Humberstone Road station was utilised for southbound trains. Swain Street staircases were dispensed with c1918 to form a closed station when ticket collectors were installed, initially on the platforms, and later in the booking hall. This practice obtained for over sixty years before the station once more became open with tickets examined or collected on the trains.

The Sabbath day was created as a day of rest, and the majority of locomotives obeyed that precept, but railway enthusiasts taking advantage of the situation would find delight in invading their privacy armed with notebooks and cameras. A view from the Birdcage of Leicester Midland yard in 1958 catches a group of slumbering engines representing the work of six different locomotive designers.

One of the few new classes to be built by the LMS between the end of World War II and nationalisation was this lightweight passenger tank engine for local duties. Here 42183 is seen working a Nottingham to London Road stopper in 1953 past Leicester North Junction signal box. The old private warehouses were demolished c1960 when a new covered goods station was erected as an extension to the original Queen Street buildings.

This March 1962 scene shows the northbound ticket platform of 1892 vintage, near Welford Road bridge. Though long disused for its intended purpose, it came in handy for railway photographers. Some local services were still steam hauled at that time, including this early evening Leicester to Wellingborough stopping train headed by 46404.

Wigston once was an important junction with extensive sidings, a locomotive depot and a wagon repair works. In 1955 one of British Railways magnificent class 9 freight engines, 92055 when brand new, heads a coal train, with two bogie rail transporters behind the tender, past Wigston South junction signal box and the Midland Terrace of former company houses.

Right: Part of the substantial 1902 station buildings at Wigston Magna. A regular working for many years was a Toton to Brent coal haul which stopped just north of the station around lunch time to take on water and to clean the fire. Getting to grips with moving the heavy train, in October 1955, is 47984 an LMS designed Beyer Garratt comprising two engines with a single boiler fed by coal stored in a rotary bunker.

47984

Leicester Central

In his excellent booklet *Two and a Half to Skegness* Tom Hammond makes no secret of his dislike of Leicester's former Central station, a sentiment which I share wholeheartedly. The booking hall was dull and uninviting, the subterranean white glazed tile passage with its signboard To The Trains, had a lavatorial air about it and the bleak windswept platforms no sense of atmosphere as did the magical Midland station.

Before her marriage my mother worked on the Great Central Railway for several years in the wages office. Her window overlooked the turntable and each November the eleventh she would recount the custom of the Manchester express, with the war memorial engine *Valour,* to observe at eleven o'clock the two minutes silence while standing at the platform on early Armistice Days after World War I. During World War II the gloom of the station matched my spirits as I set off from there, on a free travel warrant by courtesy of His Majesty, to join the Royal Navy.

My earliest acquaintance with the locomotives of the London & North Eastern Railway, which absorbed the Great Central in 1923, came, strange to relate, during history lessons at school. The Jewry Wall site was being excavated for Roman remains and our tutor would walk the class down to the site every week to watch the progress. However some of us were far more interested in the twentieth century steam trains passing by than first century architectural relics.

As a change from London Road or Swain Street the youthful train watchers sometimes met on the footpath alongside the railway running parallel to Western Road. Here a fine viewpoint was afforded opposite the engine shed. However it all paled in comparison with the set-up on the other side of the city. Unlike the quadruple tracked LMS with junctions at Syston, Knighton and Wigston the LNER was only a double track straight through route with no local junctions whatsoever. Despite the station, opened in 1899, having six platforms, compared with four at London Road, passenger trains seemed few and far between, though some carried exotic destination boards such as Newcastle to Swansea, York to Bournemouth, or Aberdeen to Penzance. However freight traffic, mainly coal, was more abundant and would rattle past Leicester South Goods signal box at a spanking forty to fifty miles an hour whereas their contemporaries on the Midland line would find themselves hard put to get up to fifteen between frequent signal stops within the city limits.

From the footpath vantage we could also see trains on the Burton-on-Trent to Knighton Junction branch of the former Midland Railway where it was carried over the Central line. In the

A Nottingham Victoria to Marylebone train sets out from Leicester Central over the canal viaduct near West Bridge in August 1966 with a scruffy 44848 at the head. Only a few weeks later London services and steam locomotives were banished from this scene for ever. The original Leicester & Swannington Railway terminus of 1832 was alongside the canal where stands the derelict building on the left.

Former London & North Eastern Railway locomotive 60830 approaches the Central station over Northgate Street bridge, past Leicester North Passenger signal box, on a Newcastle to Swansea working in August 1961.

late evening these included express beer trains on their way to Somers Town yard in the capital.

The well equipped engine shed of four roads never had an allocaton of much more than twenty locomotives. Even so during the 1950s it could boast of having on its books one of the most famous and best loved engines of all time, none other than *Flying Scotsman*. So infatuated had I become, since a toddler, with Midland Railway and London & North Western Railway built locomotives, and the rich warm crimson lake carriages, that the unfamiliar outlines of Great Central, Great Northern and London & North Eastern engines with, to me, unattractive varnished teak coaches never endeared me overmuch to The Last Main Line. Later I did come to recognise that certain classes of GCR passenger locomotives were extremely good looking.

Locomotives of the Great Western Railway were able to run onto the GCR at Banbury and made almost daily incursions to Leicester, but very rarely beyond. Once on the occasion of a Wales v Scotland rugby match at least twelve Welsh supporters' specials exchanged their GWR locomotives at Leicester for fresh LNER ones.

Some ten years after the railways were nationalised in 1948 that part of the Eastern Region which embraced Leicester Central was transferred to the London Midland Region and from thereon a process of running down took place. Withdrawal of the London, Marylebone to Manchester expresses came first in 1961, then freight trains in 1965 and the London to Nottingham service in the following year. The final demise came about in 1969 with the closure of the Rugby to Nottingham section.

Locomotive 61435 makes a spectacular effort whilst starting off with a long rake of empty stock in the down carriage sidings at Leicester Central station, This was an unusual locomotive for such a roster, normally seen only on a daily fast goods working from York.

As the Great Central was engineered to the continental loading gauge, much larger locomotives were allowed on it by the Civil Engineer's Department than on the Midland line. The 1964 East Midlander rail tour of the Railway Correspondence and Travel Society from Nottingham to Eastleigh in Hampshire, was hauled to Didcot, and back from Swindon, by Duchess class City of Nottingham. *Here it pauses on the return journey to set down Leicester passengers.*

The few bystanders on Central station, on a cold January night in 1965, viewed with suspicion the photographer who placed his camera on the slabs of the platform and then lay down to see that he had the Grimsby to West of England fish train in the viewfinder.

Steam locomotives had an insatiable appetite for water. Water tanks and cranes were a most prominent feature of Leicester Central station. Here on an August Saturday afternoon in 1961, driver John Barrow and his fireman apply themselves to topping up tank engine 42437 while on a Nottingham to Rugby working.

Graythwaite Hall *leaving Leicester in August 1960. Soon after opening the last main line to London in 1899 the Great Central Railway built a branch from Culworth Junction to Banbury connecting there with the Great Western Railway. This allowed access to Leicester for GWR locomotives on cross country services. Latterly the Bournemouth to York train was in the hands of Hall class locomotives which returned to Banbury working an early evening stopping train as far as Woodford Halse.*

Here standing at Leicester Central south bay are two highly contrasting locomotives, a spectacle which once would have been unthinkable. On the left is a former London Midland & Scottish Railway passenger tank engine, employed on station pilot duty and alongside it Clun Castle *of the Western Region awaiting the return of an Ian Allan rail tour from Doncaster on April 7, 1964. The Castle worked the train forward to London.*

On days when Jack Frost was about steam did not evaporate as rapidly as on a summer day and for truly evocative pictures such conditions could seldom be bettered. This atmospheric scene at the south end of Leicester Central station, in February 1963, shows the daily Nottingham to London Marylebone parcels train moving off in charge of rebuilt Scot 46163. Alongside, British Railways built 73000 awaits a gentle perambulation to Woodford Halse with a stopping train.

Striding across the western side of the city centre a mile and a half long series of embankments, viaducts and girder bridges carried the Great Central Railway through Leicester. On the day before the line was abandoned south of Rugby, and steam eliminated, 44847 with a short train from Rugby to Nottingham Victoria crosses the canal near West Bridge. The early Victorian hosiery factory, centre, witnessed the coming of the railway and also its demise.

A once proud rebuilt Scot displaced by diesel power from the West Coast main line. Shorn of its nameplates and relegated to Nottingham-Marylebone semi-fast services it stands in Leicester Central shed. When facilities for repairing locomotives here were suspended a fifty mile round trip to and from Leicester Midland depot by way of Nottingham was necessary. The shed was abandoned finally in July 1964.

The London Midland Region did not expel all former LNER engines when it gained control of the Central locomotive depot. Indeed two of the handsome post-war built passenger tank engines made a temporary home there. One of them 67781 lately arrived from Lutterworth on the daily pick-up goods train in May 1960, awaits permission to proceed into the yard by the Leicester South Goods signals.

The effect of the London Midland Region's take-over of the Great Central line is well in evidence here between Leicester North Goods and Leicester South Goods signal boxes. In July 1959 a former LMS passenger tank engine, 40165, with the afternoon pick-up goods from Lutterworth stands on an adjacent track to former Midland Railway veteran 47203.

The Midland Railway's branch to Burton-on-Trent was carried across the Great Central Railway on a simple girder span bridge about one mile south of the station. Having just passed underneath it, in April 1964, is a set of LNER built coaches on a Sheffield to Bournemouth through-train whose LMS designed locomotive is getting well into its stride. Although the line appears to be quadruple track one was the goods loop and another a head shunt siding

These boys at Aylestone were so intent on their fishing that they were quite oblivious of 45267 going hard up the gradient towards Whetstone. It was heading the early evening Nottingham Victoria to London Marylebone train. Less than one month later such a scene had gone for ever and a diesel multiple unit service between Nottingham and Rugby sufficed until the Great Central had run its course of three score years and ten.

Mainly Midland

Soon after the war began in 1939 films became unattainable. What few had been available were accompanied by a list of subjects which one was forbidden to photograph. Although railway installations were included there was no mention of locomotives but after photographing one at Glen Parva sidings I was reported by a railwayman to the local police, who tracked me down to my place of employment and, unfairly I thought, confiscated the film. On another occasion, during a cycling holiday, in the Chesterfield area I was apprehended by the police near a railway line on suspicion of being a German parachutist.

The sporting exercise lacking in my schooldays — according to my dad — suddenly blossomed into a passion for cycling when a soul-mate also acquired a bicycle. Our favourite summer evening run out was to Rugby to see the magnificent Duchess class locomotives, many in wartime austerity black livery, heading the long distance West coast expresses. We also visited places much further afield in our quest for the sight of unfamiliar locomotive types.

What historic photographs might have been taken in the early forties, if it had been possible, when former Caledonian Railway and South Eastern Railway locomotives were to be seen working through Leicester Midland. Not until the war had finished was I able to take up again this most agreeable aspect of my hobby. It happened first in a most unlikely place — Hong Kong — while I was serving in the Royal Navy. Using a borrowed camera belonging to a shipmate, British built locomotives emanating from Leeds and Stafford were photographed together with some American designs on the Kowloon Canton Railway.

Progress on the railways in Britain was held up during the war years yet they were the main artery on which millions of tons of essential supplies and troop movements were carried out. Trains of twenty coaches and more were not uncommon on certain routes. The only new locomotives allowed to be built were for freight duties at home and abroad because most of the railway workshops were turning out tanks and guns. The LMS class 8 heavy freight engine of 1935 vintage was chosen as the standard War Department locomotive and built in large quantities as were two new types of freight locomotive designed specifically for the War Department. At the same time, and strange to our eyes, was the influx of hundreds of American built engines for the USA Transportation Corps which could be seen working over both the Midland and Central lines through Leicester. Once Europe was liberated these locomotives were transferred to the continent.

A major change of scene occurred soon after the cessation of

Steam galore! Despite the date being as late as Autumn 1962 over thirty steam locomotives are present in this view from Upper Kent Street, now Maidstone Road. At this time the roundhouse turntable was undergoing maintenance necessitating the sojourn of this remarkable collection of engines in the shed yard. Representatives of ten different classes can be seen; one Oliver Cromwell *has been preserved.*

Until the middle thirties all main line trains changed engines at London Road but Queen's Westminster Rifleman, *seen near Swain Street bridge in 1961, would move the Thames-Clyde Express from St Pancras right through to Leeds. From there another engine would take the train onward over the Settle & Carlisle route terminating at the now demolished station of St Enoch in Glasgow.*

hostilities when the three old familiar locomotive sheds at Leicester LMS were demolished to make way for a single large pre-cast reinforced concrete roundhouse. While the building process was carried out the former Midland Railway shed at Wigston, opened in 1873, and closed as a depot in 1934, was temporarily re-opened.

Following the completion of the roundhouse, and incidentally the nationalisation of the railways, a new mechanical coaling plant and a locomotive repair shop were erected over a five to six year period. Twenty years later the roundhouse and the coaler had disappeared completely along with the steam locomotives which they had served.

A major change too had occurred in my life when I married and set up home in Highfields. With the arrival of a family railway trips and photography were limited by financial considerations but as soon as the two boys were old enough they were initiated into the noble art of platform-end loitering. I'm happy to say that this tradition has been perpetuated by my grandsons.

Another change followed later when I took up new employment. From the office in Aylestone Park one could discern, above the noise of typewriters and woodworking machinery in the manufactory, the Doncaster pitch whistles emanating from the LNER shed and the unmistakable Swindon whistle as the Bournemouth-Newcastle train hurled through Aylestone. On the Midland the Thames-Clyde Express was heard at its appointed hour, as were coal trains plodding eastwards from Coalville on the Burton branch, and the soprano shrieks of the saddle tank engines at the

Clun Castle, *the only recorded Great Western locomotive to pass through Leicester Midland, heads south on a rail tour in March 1965. The original Midland Railway signals by the bridge were still in use. At this spot one morning in December 1935 London Road Junction signal box was completely demolished when a fast goods train passing through platform 3 collided with empty coaching stock being moved out of platform 1.*

gas works and power station. Anyone working in that same location today would hear only the very occasional train of granite aggregate from Bardon Hill or Cliffe Hill quarries rumbling across Aylestone Road bridge, and at regular intervals Inter City 125 HSTs on the main line but little else.

After moving house again, this time to Clarendon Park we were nearer to Knighton South and Knighton North junctions and a new phase of my train watching career began. At this time diesel multiple units had taken over some local services and diesel-electric locomotives were being introduced gradually on Midland main line trains. I therefore decided to record on film as much of the steam workings around Leicester as I could manage.

Having discovered that lineside walking permits were available from British Railways for this very purpose I applied for one. Twenty-five miles was the maximum total allowance. It could be divided up into shorter stretches but was not valid near stations, locomotive sheds, bridges or tunnels. Having obtained the requisite permission for seven differing locations which covered all the exits from Leicester I set about making good use of it. Each permit was valid for twelve months. On its expiration one was required to submit half a dozen photographs to prove that one was a genuine photographer using the magic card to good effect before another year's freedom to walk the line was granted.

One sensible stipulation was that the permit had to be produced on the demand of any railway official. To make sure that I was not reported as a trespasser by the signalman at Knighton South

The station carriage pilot, on a January evening in 1964, its duty for the day completed. Turntables became redundant after diesel locomotives came to rule the roost, as did steam engine sheds. Beale Street roundhouse closed in 1966 after an active life of less than twenty years. Later it housed some preserved locomotives until demolished in 1970.

A Leicester London Road to Blackpool holiday train, in August 1964 headed by Jubilee class New Brunswick, *poised for the off at platform 2.*

A summer Saturday train in 1961 leaving Knighton tunnel, with 44829 running under clear signals, makes for Hastings via Cricklewood, the Dudding Hill loop, Clapham Junction and Brighton. Before leaving school I was lucky enough to spend three holidays of four weeks duration at my grandparent's home in Newhaven. We used to travel on the Sunny South Express, via Northampton and Kensington Olympia, as far as Brighton before changing to the Seaford branch train.

Working a shift at Knighton South Junction signal box, in August 1963, signalman Doug Burge operates a block instrument while Herbert, the train reporter, stands by. Additional equipment was installed to allow the closure of Knighton North box but this scheme was never implemented.

Junction the introduction of myself and my honourable intentions seemed to be right and proper. What a happy outcome stemmed from this act. That signal box, whose predecessor would have been well known to grandpa some sixty years earlier, became for a period of four or more years my second home where many a good yarn was spun by the various men on duty and many a mug of railwayman's tea was enjoyed. The same can be said, to a lesser extent, of several other signal boxes covered by the area of my permit.

I came to have a great admiration for signalmen whose demanding and responsible job was to ensure the safe passage of scores of trains and thousands of passengers daily throughout the year. When one stops to consider the career of a railway bobby, as signalmen are often known, it is quite unique. There is no job remotely like it and may perhaps only be compared loosely with that of a lighthouse keeper, especially at night when alone in the cabin surrounded by darkness with only the lights of passing trains and voices at the other end of a telephone for company.

All manual signal boxes in the vicinity of Leicester have now been demolished, either through line closures or with the introduction, in 1986, of the Multiple Aspect Signalling system controlled by a new power box, full of electronic wizardry but little soul, just north of London Road station.

Not all my train watching hours were spent on the lineside itself, after all there were two boys to be considered and their railway education could not be neglected. Ideally situated near to home was

Leicester Power Station's entire motive power fleet huddled together in February 1971 — two fireless locomotives, left, and No.3 a conventional coal-fired saddle tank.

Freeman's Common, a large area of allotment gardens stretching from Welford Road right across the top of Knighton tunnels down to Aylestone Road. It was a tranquil haven away from the noise of road traffic, surrounded by greenery, with a footpath on the east side of the railway atop the embankment. This commanded a splendid view of the south portals of the twin tunnels, and Knighton North Junction with its attendant signal box. So pleasant were the surroundings that for a walk with one's wife it was an acceptable alternative to a more conventional promenade on Victoria Park. Sometimes, when the boys and I were engaged in speculation upon which classes of locomotive might appear next, we would see a lady sitting on the grassy bank partaking of a picnic and doing her knitting. Unfortunately the Common is now covered with roads and an industrial estate.

Trains heading west from Knighton Junction along the Burton line soon passed the only two substantial industrial railways in the city of Leicester. Both of these were originally employed by Corporation establishments, the gas and electricity works. Coal gas was first produced in Leicester at the Belgrave Gate works but as the town grew a larger site had to be sought. A new works was built on Aylestone Road alongside the railway line and went into production in 1878. Steam locomotives were used from the outset for shunting coal wagons into the retort houses and around the chemical works where creosote, ammonia and carbolic acid were

Mars II *propelling a rake of sixteen-ton coal wagons into one of the old retort houses at Leicester Gas Works in July 1966.*

A signalman's eye view of the spectacular display put up by 45209 on an express fish train passing Knighton South Junction in March 1965. The south curve of the branch to Burton-on-Trent is hidden by the exhaust. Photographed from the signal box which, along with the cluster of semaphore signals, became defunct upon the introduction of colour light signalling in 1986.

A Burton-on-Trent to Leicester train, free-wheeling towards Knighton North Junction in August 1962. Saffron Lane originally crossed the Midland Railway branch on the level at this point. After the road was lowered in the early thirties, to pass underneath the railway, this signal box was retained as a day box to control the connections to the gas works and power station sidings.

Wigston Glen Parva, situated immediately behind the bridge arch, was one of three stations in the Wigston conurbation; the others being Wigston South and Wigston Magna, all now demolished. During 1986 South Wigston, the first new station in Leicestershire since 1907, was opened and is located just where 40447, with this Nuneaton to Leicester stopping train, is seen passing in July 1952.

among the many by-products. A total of eight small saddle tank engines had been purchased over the years, latterly adding one diesel, but with the arrival of natural gas, coal gas production ceased in 1969 and the sole remaining steamer became redundant.

Within the gas works complex an electricity generating station, to supply the centre of the town, was built in 1894. It was closed down when a new power station, built on Freeman's Meadow, entered service in 1922. This also was situated adjacent to the Burton railway line on the opposite side of the tracks to the gas works. Although close to the railway, coal originally arrived in canal boats which were unloaded at the power station wharf into the undertakings' own railway wagons. This water borne traffic was short lived, ceasing when sidings were laid down alongside the LMS line and connected to the power station yard via a descending incline. Two fireless locomotives only were employed until 1950 when a new conventional saddle tank arrived. The fireless locomotives were charged with high pressure steam from the station's boilers and operated happily, like big vacuum flasks on wheels, for hours before the need for replenishing arose. They were popular in environments where it was dangerous to have locomotives emitting sparks, or dropping hot ash, and very useful at establishments where high pressure steam was on tap.

With the aid of my lineside permit I was entitled to walk part way along the Burton line and could thus photograph locomotives at work on British Railways, the gas works and the power station all from the one location.

The south curve of Wigston triangle, opened in 1872, has never seen regular passenger traffic except during the decade following its opening when the Midland Railway ran a through coach from St Pancras to Birmingham. This May 1961 photograph shows 45649 Hawkins *re-starting from a signal stop with a heavily loaded Hinckley to St Pancras FA Cup Final special. On the left is the former Wigston Magna cattle dock.*

Coming off the south curve into Wigston Magna station is the Leicester Railway Society's twenty-first anniversary special. The bridge was built in 1902 to replace a former level crossing when the tracks were widened and imposing station buildings erected. Passenger services were withdrawn on January 1, 1968 and the station subsequently demolished.

Belgrave Road

Leicester's second station of main line status, though in reality the terminus of a ten mile branch line, was the pretentious overall-roofed mini Kings Cross in Belgrave Road. It was officially opened to passenger traffic in 1883 although one excursion train to Skegness had left the station the previous year, the same in which goods trains had commenced using the adjoining freight yards. The provision of six platforms at Belgrave Road was ambitious in the extreme. They were never justified by the sparse passenger traffic. However the freight side was a different and more successful story outliving the passenger trains by almost seven years. A considerable volume of varied goods traffic was handled including, of course, coal.

Leicester's tramway system was electrified in 1904, powered from a generating plant built in Painter Street just across Belgrave Road from the railway. Coal to feed the boilers was transported from the railway goods yard to the power station by road steam wagons. During the miners' strike of 1926 conversion to oil firing took place and railborne oil fuel was transferred to the tramway's brine tank cars for the short journey across to Painter Street.

Belgrave Road station is remembered with affection as the departure point for the seaside and it was, I suspect, ironic but fitting that the last public train to use the station, in 1962, was an excursion to Skegness just as had been the very first train nigh on eighty years previously. The line remained faithful to steam throughout and only after closure were diesels to be seen, performing the unhappy task of transporting away the lifted track.

The goods warehouse, latterly in private hands, had the last laugh for it was not demolished until 1985, thirteen years after the station had suffered the same fate.

My earliest recollection of seeing this railway was when dad used to take my brother and I to the annual May Fair held on waste ground in Ulverscroft Road. Later there was a short period when I saw probably more of the former Great Northern Railway's installation between Belgrave Road and Forest Road than most; excepting of course railway employees.

It was in 1940, after the collapse of France and the evacuation of the British armies, that a part time civilian army known as the Local Defence Volunteers was formed. This title was soon given a sharper image when it became the Home Guard. Not yet being eligible for call to the armed forces I signed on, at Charles Street police station, for membership of the LDV. Imagine my surprise and delight when I was summoned to report to Belgrave Road station!

During training we had to attend parades every Sunday morning

Belgrave Road station closed in September 1962. A year earlier it was still very much alive though only during the summer months. Here 61163 is seen recuperating after jogging all the way from Mablethorpe. The noble spans of the train shed were not matched by the rather plain exterior of the building. Ten years after the last timetabled train had run the station carcase was finally removed.

There appears to have been an overabundance of coaching stock in and around Belgrave Road station on the last day of the 1961 summer timetable when this day trip departed for Skegness. Many people thought that this would be the very last train but a later reprieve meant that there was a repeat performance the following season. It must have been a highly unecomonical operation as the locomotive had to travel light from Colwick depot at Nottingham in the morning and return there from Leicester at the end of the day.

Leaving Leicester at ten minutes past five on May 29, 1964 the last revenue earning train ran on the Marefield Junction to Belgrave Road branch. Locomotive 61361 with the final daily goods train to Colwick, passing a burned out platelayers hut, and an already disconnected GNR signal, sets off to tackle the climb to Humberstone. On the left is the chord line to the Midland which came into use on the following Monday.

when one of the platforms would become our barrack square. The marching feet of volunteers had replaced the lighthearted steps of holiday makers and scurrying factory workers. Night guard duties would consist of patrols around the goods shed or along the track as far as Forest Road level crossing or guarding Catherine Street bridge. Sadly there were no trains during the night watches, though we did witness once an evacuee special from the East coast. Orders forbade anyone from trespassing near the small three-road engine shed but a certain volunteer who had a passion for steam locomotives would often have a quick look around the interior during the small hours.

Incidentally it was under cover of the wartime blackout in 1940 that I had my first footplate ride on a locomotive, travelling from Glen Parva into Leicester.

There was a Great Northern flavour present in this 1959 August Bank Holiday scene just outside Belgrave Road station, where 61754 is waiting to push a rake of empty coaches under Catherine Street bridge to form the 11.05am day excursion to Skegness. This locomotive and the lattice post somersault signal were both built by the GNR.

After the withdrawal of freight services a new connection was made between Belgrave Road goods yard and the former Midland Railway along the embankment used by the contractor building the GNR in the 1880s. Seen descending the new line near Forest Road in November 1964 this empty train formed the Leicester Railway Society's special for a historic trip from platform 6 of Belgrave Road station to West Bridge.

The summer only weekend services from Belgrave Road to Lincolnshire coastal resorts were often made up of twelve coaches. Here one such train, in July 1961, is facing the gruelling gradient through the Leicester suburbs. With a rake of LMS built carriages 61227 is leaving the overbridge carrying the Midland main line and crossing Forest Road on the level.

In the year before passenger trains finally ceased on this line a train for Skegness pulls into Humberstone station to pick up holidaymakers.

Leicester & Swannington

The Leicester Railway Society after its modest conception in 1939 with a handful of recruits grew rapidly in the fifties and at its peak had a total of one hundred and twenty four members which included a branch in Nottingham. In those days railway enthusiast societies were unheard of apart from the few principal ones, namely the Railway Correspondence and Travel Society, the Stephenson Locomotive Society, the Birmingham Locomotive Club and the exclusive Railway Club in London. Nowadays with the tremendous upsurge of lay interest in railways almost every town and city has one or more such organisations added to which are the various branches and support groups of preservation railways.

Unlike some groups the Leicester Railway Society organised few rail tours. However, those which they did promote were of historic interest and in retrospect are now even more so. Twice, in separate years, members were allowed to travel in a brake van of the daily goods train on the former Leicester & Swannington Railway line between West Bridge, Leicester, and Coalville via the Glenfield tunnel and Desford Junction.

On the occasion of their twenty-first anniversary, albeit almost a year late due to organisation difficulties, a special charter train made a tour of branch and freight-only lines in Leicestershire. This included the Leicester & Swannington line again from Desford Junction to Glenfield and back. The train was not able to proceed through Glenfield tunnel as there were no passenger coaches remaining which could clear its narrow bore.

The tunnel was a remarkable achievement in its day, engineered by Robert Stephenson and opened in 1832, one mile thirty-six yards in length. During its construction geological problems were met and due to the extra costs involved the planned Mountsorrel granite portals at the tunnel mouths were never erected. The line, which was the first steam railway in the Midlands, was built solely for the carriage of coal into Leicester. Large passenger coaches and locomotives were never envisaged when the tunnel was built so throughout the whole of its one hundred and thirty-four years of service there were restrictions placed on the size of rolling stock.

The first of two stations built at the West Bridge terminus was situated alongside the canal. Passengers were of secondary consideration and conveyed in open trucks attached to coal trains until 1887. Furthermore a passenger platform was not provided until 1876. From 1848 the importance of the line had been lessened with the opening of a new route into Leicester from Desford Junction to Knighton Junction by the Midland Railway which had

West Bridge sidings, in July 1965, with engine 78028 waiting to take the afternoon goods train to Desford junction. Petrol tankers, bulk grain and coal wagons were still in evidence. Eight months later this locomotive had the sad duty of hauling the very last train out of this historic yard. The two ancient churches of St Nicholas and St Mary de Castro, which preceded the railway by over seven hundred years, stand sentinel on the skyline.

West Bridge wharf was extended in 1834 by bridging the river Soar and laying on its eastern side extensive horse worked sidings known as Pingle Wharf. Robert Stephenson, the engineer for the railway, designed this wooden bridge here seen disused in 1963. A section of the span was arranged to lift up to allow boats to pass beneath.

purchased the Leicester & Swannington four years earlier. A new station at West Bridge, built nearer to Tudor Road than its predecessor, opened in 1893; its single platform and plain brick structure contrasted starkly with the contemporaneous station at London Road. West Bridge remained in use for passengers until 1928 when the service along the branch to Desford was withdrawn. The line remained open for freight and continued to serve West Bridge wharf with its coal and oil facilities until 1966 when the curtain finally fell witnessed by only 'three men and a boy' to quote one report.

The Leicester Railway Society's final excursion along the line, in celebration of its 25th anniversary was in 1964. On this occasion a unique journey to West Bridge platform from Belgrave Road station was made possible by the opening of a new connection, from the Midland line at Humberstone Road junction, to the former Great Northern line to serve the goods depot. Hauling the train of brake vans was 78028, a modern locomotive which, along with another of the same class, had been specially adapted in 1963 to negotiate Glenfield tunnel. The three surviving eighty year old Midland Railway built locomotives then were retired and consigned to the scrapyard.

This locomotive also had the distinction of unceremoniously moving the last train out of West Bridge yard, with sorrow among the staff, in contrast to the gala celebrations and joyous pealing of bells which marked the departure of the first train with so much enthusiasm on a summer's day in 1832.

Thus far and no farther can the Leicester Railway Society special of May 1961 proceed along the branch to West Bridge because of the small bore of Glenfield Tunnel. The three coach push-and-pull set standing proud of Glenfield station awaits the return of passengers inspecting the mouth of the one mile long tunnel before returning to Desford Junction. The train then ran to Coalville and on to the Charnwood Forest line. Later that afternoon it became the first, and only, steam passenger train to traverse the Enderby Quarry branch from Narborough.

Passed Fireman Ellis Hill takes over the controls for a spot of shunting at Desford Junction transfer sidings.

Before taking his class 8 freight locomotive and short train onward to Knighton Junction in February 1965 Driver Harry Johnson, of Coalville depot, walks towards Desford Junction signal box for a quick word with the duty signalman. The straight track to the left of the box passed through the disused stations of Ratby and Glenfield on its way to the terminus of the former Leicester & Swannington line at West Bridge.

Local Stations

At the turn of the century there were thirteen passenger stations within a four mile radius of Leicester's Clock Tower. Apart from the withdrawal of passenger facilities at West Bridge and Glenfield, in 1928, this remained the case until all but London Road were swept away during the so called Beeching era of the nineteen sixties.

Except for Leicester's principal stations all were of the suburban or country variety and in the years before mass personal transport handled hundreds of commuters daily. They were also useful revenue earners as feeders to the main line stations for long distance travellers. The majority had a station master whose house was in some cases part of the station itself. The staff could include porters and passenger clerks, often local people themselves and known by name to many of the travellers. Parcels traffic was just as common as passengers on local trains and often one could see horse boxes, cattle trucks or milk tanks coupled onto the coaches.

The stations themselves were as varied as the types of passengers who used them. The London & North Western Railway's simple offering at Glen Parva had a wooden ticket office at street level which matched the austere shelters on the platforms. In contrast the Midland Railway's splendid building at Wigston Magna had an ornate frontage adorned with terra-cotta, in the company's best architectural style. In between these extremes was a wide variety of brick built and timber structures, no two alike. Blaby station, for instance, stood on an embankment and had all the appearance of a halt.

The once common sight of racing pigeons in flat wicker baskets being loaded into the guards van of a passenger train for conveyance to their designated starting points was always an interesting spectacle. A railway employee, who later in life held a managerial post on British Railways at Leicester, confessed that when young and inexperienced he released several baskets of pigeons from the station at which he then worked not knowing that they had just arrived for onward transit to a distant starting point for the race home. The pigeons reached their lofts, nearby, before their irate owner!

Harking back to my Alderman Newton's schooldays there were two lads in my form who lived in Birstall and went home to lunch. To enable them to do this they were allowed to leave five minutes before the end of the last lesson of the morning so that they could catch a train from the nearby Central station to Belgrave & Birstall. Their return journey also was made by rail and they were always back in good time for afternoon school. Another boy in my class lived at Ingarsby where his father was the station master.

Humberstone Road station, although only threequarters of a mile north of London Road, was no more than a wayside station. The wooden raked palings were a typical Midland Railway feature. Beyond is J.O. Walker's timber yard. On a Spring evening in 1960 this stopping train from Leicester makes a lively getaway, with 44509, on its way to Peterborough. A year or two later this train was often hauled by Britannia class Pacifics with power to spare.

Thurnby and Scraptoft station, a simple brick structure with cantilevered canopy, was equidistant from the villages which it served from 1883 onwards. By 1951 there was only one train a day remaining in each direction and this service was withdrawn two years later. However an unadvertised workmen's train from John O'Gaunt to Belgrave Road continued to run until April 1957. After that date only passenger excursions and goods trains used the line until they too were withdrawn.

Notwithstanding the sparse passenger service on the former Great Northern line he was able to travel into, and home from, Belgrave Road station daily.

When I worked in South Wigston the train journey from Leicester took a mere six or seven minutes. My chief, at that time, who lived in Countesthorpe, often travelled home for lunch, taking a train from Wigston South. He was back in the office within an hour and a half, such was the convenience of the service. It was not that there was a train every few minutes, but those that ran were timetabled to suit factory and office workers at peak times.

In addition to the regular daily services the country stations were served also by various categories of excursion trains and privately chartered specials including works outings and Sunday School trips. Such trains which once called at these stations are redolent of a bygone era. Hauled by a steam locomotive, of whatever size or age, they were an unforgetable sight, fair to behold.

A Rugby to Leicester stopping train arrives at the staggered platform of Wigston South on the former Midland Counties Railway route to London. The architecture of the signal box, footbridge and adjacent signal is typical of the Midland Railway. Four months later the same locomotive, 42331, headed the Leicester Railway Society's last day farewell special over the line on December 30, 1961.

There was no doubting when a Great Northern signal was in the clear position as the arm completely divorced itself from the spectacle and post. This one at Humberstone station had possibly survived since 1882 when the line was opened to freight traffic. All this scene has long gone, including the bridge over Uppingham Road, and the railway embankment.

The standard cardboard railway ticket, measuring 2¼×1 3/16 inches, was introduced by Thomas Edmondson, c1838, who patented his first printing machine in 1840. This style of ticket was widely used and it was not until December 1987 that the last tickets of this type were printed for British Rail, whilst some preservation railways continue to use them. Now and again mistakes occurred in the spelling of names and that some featured quite superfluous information. Two examples of the former, Ingersby and Enderly, appear in the selection of local tickets, shown. The alternative route to 'via direct' on the Humberstone Road to London Road ticket would have been somewhat more lengthy by way of Melton Mowbray, Oakham and Kettering!

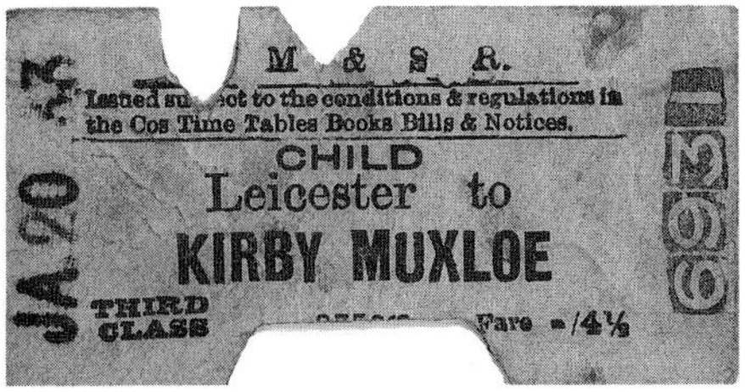

Child's third class ticket to Kirby Muxloe, 1933, fare 4½d.

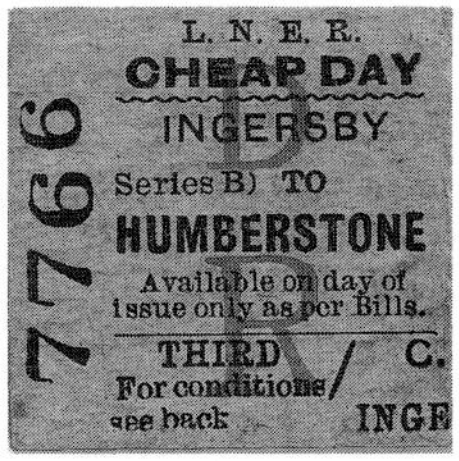

Ingarsby is mis-spelt Ingersby.

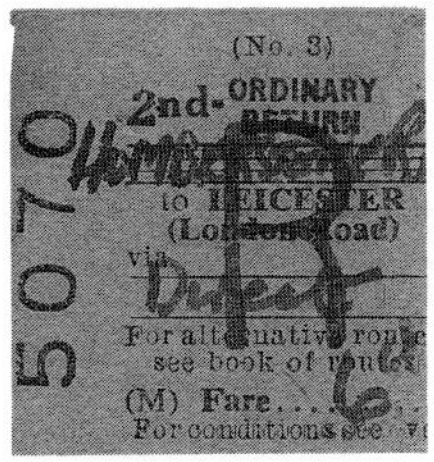

Humberstone Road to London Road.

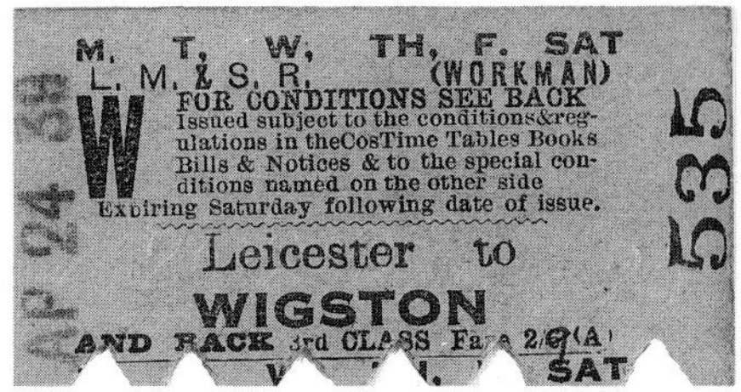

Workman's weekly season ticket issued 1939.

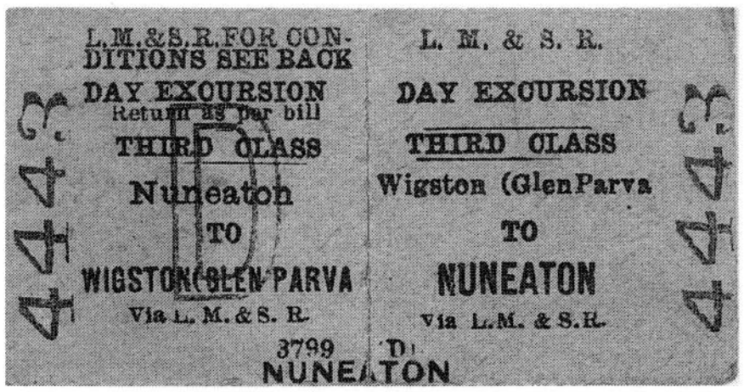

Day excursion, Nuneaton to Wigston Glen Parva, 1940.

A Leicester Central ticket.

From West Bridge to Desford via Daily Goods Train, 1958.

Leicester Railway Society's 21st Anniversary Special.

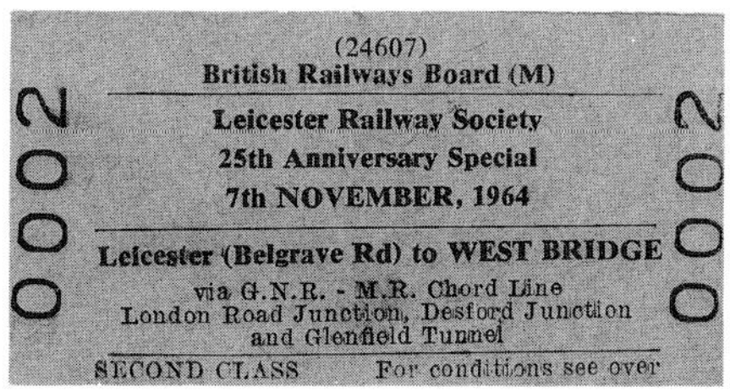

LRS charter train from Belgrave Road to West Bridge, 1964.

The porter at Belgrave & Birstall who had just warned me quite seriously that no photography was allowed on his station, walks away to attend to any passengers who might alight from the Leicester Central to Nottingham stopping train in August 1959. The employment on this line of these former LMS suburban tank engines, 42339 among them, was short lived. The station was closed in March 1963, and the building later pulled down.